Published by Gluten-Free by Jan LLC

Gluten-Free by Jan LLC
16067 NW Rondos Dr.
Portland, OR USA 97229-9239

Counting in Tagalog Copyright © 2026

Text and Illustrations Copyright © Jeanette Withington 2026

ISBN: 979-8-9945334-0-6

Counting in Tagalog

Let's count together!

Isa, Dalawa, Tatlo

Jeanette Withington
&
Edith Withington

Isa – One
(ee-sah)

Isa - One

(ee-sah)

Isa, number one,
One bright sun, hello sun!
Point up high and shout with glee,
Isa! Count with me!

Dalawa - Two
(dah–lah–wah)

Dalawa – Two

(dah-lah-wah)

Dalawa, number two,
Two little hands, me and you!
Clap them once, clap them tight!
Dalawa! That feels right!

Tatlo – Three
(taht-loh)

Tatlo – Three

(taht-loh)

Tatlo, number three,
Three small birds in a tree
Isa, dalawa, tatlo, watch them
go,
Tatlo, flying low!

3

Apat – Four
(ah-paht)

4

Apat – Four

(ah–paht)

Apat, number four,
Four rolling wheels on the floor,
Round and round, fast they roll,
Apat! That's our goal!

Lima – Five
(lee-mah)

Lima – Five

(lee-mah)

Lima, number five,
Five wiggly fingers, wiggle alive!
Wave hello, wave some more,
Lima! Count and explore!

Anim - Six
(ah-nim)

Anim – Six

(ah–nim)

Anim, number six,
Six hopping frogs do funny tricks.
Jump up high, jump down low,
Anim! Go, go, go!

Pito – Seven
(pee-toh)

Pito – Seven

(pee-toh)

Pito, number seven,
Seven stars shining up high.
Twinkle, sparkle, night-time glow,
Pito! Soft and slow!

Walo – Eight

(wah-loh)

Walo – Eight

(wah–loh)

Walo, number eight,
Eight fishes swim, wiggle and
wait.
Swish and splash, tails swish by.
Walo! In the bright blue sea!

Siyam – Nine

(see–yahm)

Siyam, number nine,
Nine balloons, oh so fine!
Up they float, high and far.
Siyam! You're a counting star!

Sampu – Ten
(sahm-poo)

10

Sampu - Ten

(sahm-poo)

Sampu, number ten,
Ten little ducks, count them again!
Wiggle and waddle, quickity-quack.
One more duck goes flap-flap-flap!
What a sight!
Sampu! You're a counting star!

Let's count again!

1. isa – (ee-sah)
2. dalawa – (dah-lah-wah)
3. tatlo – (taht-loh)
4. apat – (ah-paht)
5. lima – (lee-mah)
6. anim – (ah-nim)
7. pito – (pee-toh)
8. walo – (wah-loh)
9. siyam – (see-yahm)
10. sampu – (sahm-poo)

Great Job!

You counted Isa hanggang Sampu,
In Tagalog, yes you do!
Count again fast or slow,
Mabuhay! Way to go!

hanggang – up to
(hahng-gahng)

mabuhay
(mah-boo-high)

www.ingramcontent.com/pod-product-compliance
Lightning Source LLC
Chambersburg PA
CBRC090842120626
46551CB00008B/729